DISCARDED

E591.9 Zingg, Eduard, 1940-
ZIN $14.90
c.1   Animals of the Delta

~~~ - 6 1994

DEC - 7 1994

Driftwood Library of Lincoln City
801 S.W. Highway 101
Lincoln City, Oregon 97367

ANIMALS OF THE DELTA

Written by Eduard Zingg

Published by Abdo & Daughters,,6535 Cecilia Circle, Edina, Minnesota 55439

Copyright © 1993 by World Wild Life Films (Pty.) Limited, Postfach 6586, 8023 Zurich, Switzerland

Edited By: Jim Abdo and Bob Italia for Abdo & Daughters Publishing

Text and Photographs: Eduard Zingg
Illustrations and Maps: W. Michel and K. Wozniak

International copyrights reserved in all countries. No part of this book may be reproduced without written permission from the copyright holder. Printed in the U.S.A.

Library of Congress Cataloging-in-Publication Data

Zingg, Eduard, 1940-
  Animals of the Delta / written by Eduard Zingg: [edited by Bob Italia].
    p. cm. -- (An African animal adventure)
  Includes index.
  Summary: Follows an expedition through the Okavango Delta region in Botswana and describes some of the different animals found there.
  ISBN 1-56239-217-4
  1. Z oology -- Botswana -- Okavango River Delta -- Juvenile literature. [1. Zoology -- Botswana. 2. Botswana -- Description and travel.]
I. Italia, Robert, 1955- . II. Title. III. Series: Zingg, Eduard, 1940- African animal adventure.
QL337.B68Z56 1993
591.96883 -- dc20   93-3700
             CIP
             AC

Driftwood Library
of Lincoln City
801 S.W. Highway 101, #201
Lincoln City, Oregon 97367

E591.9
ZIN
c.1
ENF

Diamond Lake Book Co. 14⁹⁰   3-30-94

Table of Contents

PARADISE IN THE WILDERNESS

After a good sleep we were up early in the morning to prepare for our trip into the bush. We knew that the next 125 miles (201 kilometers) would be hard going, but we were ready for it.

Towards midday we reached a village of the river bushmen. We parked our trucks and got out. Mutero, my assistant, helped us by speaking their language. Armed with the rifle, camera and sweets, we walked through marshy terrain to get to the village. The villagers welcomed us.

The river bushmen live on the fish they catch and whatever they can hunt. Their clothing is made of skins and they hunt with spears. It was a lucky thing that the people happened to be in the village. Mutero explained that they had come from the north of the swamps because the water had risen there and flooded the dry land on which they lived. We filmed the bushmen and recorded their music. Towards evening we went back to the trucks where our friend Joseph had been waiting since midday. After an interesting day, we went to sleep on our camp beds under the stars.

The following day we encountered very hard traveling. We intended to travel southwest, but sometimes I was not quite sure if we were heading in the right direction. I knew this area well but it was hard to find the tracks. Often we had to go through water 165 feet (50 kilometers) across. We were always happy to pick up the track on the other side. What impressed us the most on this trip was all the sable antelope which were in the area. I pointed out to our friends Robert and Peter how fascinating these animals are. We went on foot as close to them as we could.

AFRICAN ANTELOPES

Orygine (or-E-gin-ee) antelopes hooves point sideways (lateral hooves). Both females and males have horns. They have hairy muzzles and no face glands. Both the sable antelope and the roan are members of this family.

Whether the sable or the kudu bull wins the prize for grace and beauty is largely a matter of opinion. There is a certain fearlessness and pride in the sable which isn't in the other antelopes. Its huge horns and the brilliant contrasts of rich brown, black and pure white in the coat render it the superior animal in many people's minds.

The sable antelope, with its satiny coat and its strong body, reminds me of a full-grown Arabian stallion. Adult bulls are very dark brown to almost glossy black above and on the neck and limbs. They are pure white on the belly. There are white tufts of hair in front of the large dark eyes. The chin is also white.

The sable antelope, with its black coloring, satiny coat, and gracious physique, reminds one of a full-grown Arabian stallion.

The ears are narrow and resemble a horse. Small calves are fawn color with distinct facial markings. The winter coat is rough and less glossy. The tail is long and fringed on the sides. The 40 inch (1 meter) horns are heavily ridged, stout and swept back. At the shoulder the sable stands 4 feet (1.3 meters) and weighs an average of 555 pounds (250 kilograms).

The average herd is 20 to 30. But there have been herds up to one hundred and more. The old bulls separate themselves from the herd and live alone or in pairs. They eat leaves and shoots, preferring thin forest country to live in. They are regular drinkers and are seldom found more then a few hours from water.

Sable antelope are very exclusive in their choice of company. You will seldom see them with other antelopes. The bulls like to be alone when they are drinking. They will chase away other animals that approach the waterhole. When they appear at the waterhole most of the other animals leave. The sable is very proud and courageous. The bulls fight each other with a loud clashing of horns. Their horns are dangerous weapons. No wonder other animals avoid them!

The sable bull is seldom killed by other animals except the lion. Even a pack of wild dogs will stay clear of a bull. When the lion attacks it makes sure it doesn't get too close to the sable's horns.

The sable cows and calfs are a reddish-brown color.

INTO THE DELTA

My friend Bob, the crocodile hunter, accompanied us into the delta. There were nine people in the party. We drove about 20 miles (32 kilometers) out of Maun and there, in the midst of pink, yellow, and white waterlilies, we packed up the boats. We would leave the landrovers behind.

Our next call would be on the river bushmen. Of course we would watch for animals. We were sure on this expedition we would come across lechwe, sitatunga, and crocodiles. These animals were much deeper in the delta.

To get a good head start we decided to travel through the first night. Although we had not slept well on the boats, we were rewarded with a fantastic animal scene the next morning. Many different herds of animals came to the water to drink. On the island where we stopped to have breakfast, Bob came across a huge python snake.

We travelled northward past many small islands. We enjoyed gliding along the waterways. The water was crystal clear. Our speed was slow because we had packed the boats full of supplies. This was much different than travelling in a bumpy landrover in the middle of clouds of dust. We felt as if we were in paradise, a paradise in the wilderness, untouched by man. No newspapers, no radio, and no telephone. There had never been any tourists here. They had probably been kept away by the tsetse-flies, and the fear of sleeping sickness which they cause. Perhaps some had tried but never found their way back. Without Bob I would not have ventured into the delta myself. When I told him I appreciated him coming with us, he just laughed.

An enticing shifting wilderness of water and grassland,
the Okavango Delta supports a vast range of African wildlife.

Treeless stretches of land reveal the position of recently dried waterways.

As the sun sank, the birds flew home. Other creatures became active, including the mosquitoes. Bob told me that we would soon see many animals, including crocodiles and hippos. Bob then told me of seeing a river bushman drowned by a crocodile. The crocodile tipped over his makorro (canoe), grabbed his leg and dragged him under the water. We were now approaching Chiefs Island, the main island in the Okavango Delta.

As I passed an island, I recalled a fond memory. Many years before a chief of the native bushmen tribe had given me my own island as a gift. I had never seen it because I was always there during the rainy season and it was always underwater. We decided to camp on my island for the night. It started to get very cold but most of us liked the relief from the heat. In the morning we went to Chiefs Island for filming. It was full of animals. We had hardly stepped out of the boat when we encountered a herd of elephant walking towards us. There were lechwe very close to the elephants.

Antelopes are probably the most beautiful and graceful of all wild creatures. Almost no sight can compare with that of the great herds of lechwe as they move about on the open grassland plains.

LECHWES

Lechwes (LEK-ways) are a delicate-looking type of antelope. They have a reddish-brown coat, darker along the back, with the chin and throat being white. Only the males have horns. The horns grow to about 24 inches (61 centimeters).

12

A lechwe's height at the shoulder is about 36 inches (91 centimeters). They weigh about 244 pounds (110 kilograms).

Lechwes live in swamp areas. They are found in the Okavango Delta. They are somewhat aquatic, feeding belly-deep in water. They are also excellent swimmers. They seem to have no fear of crocodiles, and they even seem to act as a warning to them and to hippos of the approach of man.

When galloping, lechwes lay their horns back on their shoulders. On dry land, the lechwe has difficulty moving around quickly. But it can gallop with ease through thick wet mud. Their hooves are about 3 inches (7.6 centimeters) long. Herds are usually small, although they may grow to a few hundred at certain times of the year.

The scenes changed one after the other. Everything seemed to fall into our laps. First we filmed elephants. Then a large herd of impala came jumping by. They seemed to dance as they arched their backs and leapt into the air. With their light colored bellies and darker backs, the impalas seemed to be hunting us. They ran into the camera head-on. They rocketed into the air and then ran away.

THE IMPALA

Impala are found wherever there is warm bush country with a good supply of water. Their enemies are the leopards, lions, wild dogs, cheetahs, and crocodiles. In general, they are too fast and alert to be caught by lions. They drink often but hardly dip their mouths into the water, as they are always on the lookout for danger. One or two will stand guard while the others drink.

The impala are beautifully proportioned, with nobly-shaped heads, big ears, slim necks, and bodies set on long slender legs.

It is great fun watching an impala. It has graceful movement and form. If you are able to observe them without being seen, you can see them staring around with their big brown eyes at the thornbush country which they prefer. You can watch and study these animals. Where tender grass grows after a brush fire they graze peacefully as their leader stands guard at the edge of the herd.

Impala live in herds of 50 to 100. The leader is usually an older male. For security the mothers and the young are usually in the center of the herd. When impalas are in large herds they feel secure. But when they are alone they are very timid. When the herd is threatened, the bush echoes with their high-pitched alarm call. Then suddenly the whole herd will bound into cover. It is fantastic to watch as they make a series of short, jerky jumps before soaring away with breathtaking leaps. They can jump as high as 10 foot (3 meters) and as far as 30 feet (9 meters). Often in their hurry to get away they appear to be sailing over one another.

Impala are beautifully proportioned, with nobly-shaped heads, big ears, slim necks, and bodies set on long slender legs. They stand about 33 inches (84 centimeters) at the shoulder and weigh about 155 pounds (70 kilograms). They have a sleek glossy coat of short reddish-brown hair. They also have gracefully curved horns inclining slightly upwards and inwards at the tips. The horns reach a length of about 24 inches (61 centimeters) and are used by the males for fighting each other. Males often wound each other badly. They make loud grunts and snarls, almost as loud as angry lions.

A few hours after a young impala is born, it runs around. Within days it joins the rest of the herd. Impala are very curious animals. It is possible to get close to them. Black tufts of hair can be seen growing behind the back foot. The hair conceals a scent gland. Scientists believe that this scent gland is used in keeping the herd together. Impala can often been seen in the company of other animals.

Once we came across an impala who seemed to be posing for a picture. One of us who was not photographing him noticed a herd across the road some distance away. As soon as the herd was out of sight our impala took off. It had been standing guard and distracting us while the herd passed in safety.

I once witnessed an encounter between a crocodile and an impala. I came out of the bush and into a clearing, near a waterhole. Drinking there was a herd of impala. They all had their noses in the water. Suddenly all the impala jumped back, except for two which remained at the water's edge. There were ripples on the water, and a crocodile came to the surface. It shot out of the water with its front legs and grabbed the first impala by the leg. The crocodile began to swing the impala from side to side. A rattle came from the poor animal's throat. I moved forward with my rifle as the rest of the herd sprang away and out of sight. I was too late. The crocodile had already dragged the impala under the water. "The law of the wilderness," I thought. Sadly, I continued on my way.

Impalas live in herds of 50 to 100, consisting of the leader, usually an older male, a few young males, and many females and young.

Driftwood Library
of Lincoln City
801 S.W. Highway 101, #201
Lincoln City, Oregon 97367

17

THE CROCODILE

The crocodile is the most unpopular creature in Africa. The crocodile crawls slowly on land, with its long jaw, heavy body, and a dragon-like tail. The crocodile's nose, head, body, and tail are continuous, with only the stumpy, crooked little legs showing where the belly begins and ends.

The crocodile feeds only on meat. Its long jaws are filled with sharp, short teeth, which are not ideal for tearing fresh meat. This is why the crocodile stores its prey underwater. There it is softened by the water. The jaws of the crocodile are very dangerous. So is the powerful tail. The tail can hit and wound anything in its way.

Crocodiles reach a length of up to 16 feet (5 meters). They lay their eggs in the sand near the riverbed. The young hatch after about 14 weeks. The newly-hatched young are often eaten by leguaans, baboons and eagles. Larger crocodiles will eat smaller ones, but parents treat their young with care. When the young hatch they are about 10 inches (25 centimeters). If they find themselves in danger, they squeak for help. After 12 months the baby crocodile is about 22 inches (56 centimeters) long. After that they grow about 12 inches (30 centimeters) a year until they are three years old. By then, they measure 40 inches (102 centimeters).

The heavy outer skin with its scaly pattern provides an armor against any attacker. The only weak spot is the eyes and a spot behind the head. The surest way to be certain that a crocodile is dead is to cut the spinal cord. Crocodiles live to be very old. Their teeth and even their limbs grow again if they are lost.

Baby crocodiles hatching from their shells.

The crocodile is able to stay underwater for long periods of time. The older it gets the longer it can stay underwater. When it is 6 feet (2 meters) long, it can stay underwater for 45 minutes. The ears of the crocodile extend backwards from the eyes. Like the nostrils they can be closed by a muscular valve on the outside of the head. Its eyes are similar to those of a cat, with a pupil that dilates with the approach of night. At night is when it sees best. This is when it does most of its hunting. Its eyesight and hearing are excellent but it has a poor sense of smell.

Apart from the tail, crocodiles are relatively harmless on land. They go ashore only to lay their eggs and bask in the sun. Their area of operation is the water where they wait for their prey to drink. Hovering near the bank just below the surface of the water, they rush out to attack. They grab the victim in their jaws and drag it into the water to where it drowns. Usually they don't eat right away. The crocodile likes its meat rotten. The prey is stored on an underwater shelf until it decays. Because of this habit, the crocodile has a foul-smelling mouth or breath. When operating in water, the crocodile will take on anything its own weight or even heavier.

Like most amphibians, a crocodile has webbed back feet. But the front foot has five toes which look like a paw or hand. The toes are similar to those of dogs, and they allow the crocodile to dig a nest for itself. Crocodiles also dig tunnels in which they hibernate over the winter period. Until the age of three years, they are shy and will run to the water if approached. But after this age they become more aggressive and will make an effort to defend themselves or even attack.

In some parts of Africa, crocodiles are protected game and can only be destroyed if they are attacking cattle or humans. Although the crocodile generally attacks its prey in the water, it is able to take animals on land. The crocodile propels itself out of the water with a powerful movement of its tail. It comes out of the water with a rush to grasp animals such as warthogs or young antelopes.

Among the many enemies of the crocodile, the most effective is the elephant. It can trample to death any crocodile unwise enough to cross its path on dry land.

The following morning presented us with beautiful scenes of the wilderness. On the shore of Chiefs Island, fish eagles were sitting in the trees drying their feathers in the sun.

It seems that the Okavango Delta is the home of the fish eagle. It lives wherever there is water for the fish which is its main diet. You can see fish eagles all over the delta. Pair after pair can be seen along the tree-fringed shore, each with its own well-defined territory. They sit for hours on end. It is a fascinating sight to see them dive down to capture a fish. It swoops feet first over the water, and grips the fish with its strongly hooked talons.

THE AFRICAN FISH EAGLE

Fish eagles have been known to kill fish that are 5 pounds (2.5 kilograms), though a victim of this size is difficult to lift. Fish eagles have attacked fish that are far too heavy. When they dig their claws deep into the fish, they can't lift it out of the water. Nor can they pull their claws out. Then they are pulled underwater and drowned. Their favorite game is to swoop down on other birds and make them drop their prey. The fish eagle then steals the prey for itself. Fish eagle have been killed playing this game against goliath heron.

The fish eagle is never far from water. It has vivid colorings. The head, breast and tail are pure white. The back and wings are mainly black. Wing covers above and below are red while the cheeks and legs are yellow. It has a large hooked bill and heavy claws or talons. The fish eagle's thrilling cry may be recognized immediately. Their call is one of the most memorable sounds of the African bush.

The African fish eagle is a versatile predator, taking the young of flamingoes, as well as snatching fish from close to the surface of pools and streams.

The nest is typical of large eagles. It has a platform of sticks high up in a tree. When the eggs are to be laid the nest is lined with green leaves.

Another riverside dweller is the waterbuck. This is an animal of large or medium size, with well-developed horns which are curved forward. Waterbuck are among the most handsome of the larger African antelope.

THE WATERBUCK

The common waterbuck has a white ring around the rump at the base of the short tail. There are also white areas around the eyes and nostrils. This animal is found mostly in southern Africa.

Waterbuck have a coarse, grayish-brown coat. There is shaggy hair around the neck, and the heavily ringed horns are up to 3 feet (1 meter) in length. The waterbuck stands up to 3 feet (1 meter) at the shoulder. It weighs about 525 pounds (235 kilograms). The ears are short and broad.

Waterbuck are among the few antelopes which can eat the rough, reedy weeds near the water. Usually the herds number about twelve, although they can reach thirty in size. The cows and the bulls live apart most of the year. The calves are usually born in the summer months.

The common waterbuck has a long coarse coat,
which seems best adapted to the terrain in which they live.

Although they live near the water, they sometimes wander into the bush and across the plains. They always make their way back to the water after a short period. They are excellent swimmers. They know how to hide among the reeds when danger is near. If they are being chased and cannot escape they take to the water. They disappear under the water until only their nostrils are visible. Sometimes they will even swim across deep rivers. This means their enemies can only approach them by swimming. This gives the waterbuck a great advantage. Fear of crocodiles usually prevents the enemy from following the waterbuck into the water.

Most antelope stand back from the water in a cautious stance when drinking. The waterbuck goes right into the water, often above the knees. They take their time when drinking, satisfying their thirst. They do not seem to have any fear of crocodiles. This may be because of the sharp odor they emit. The smell of the waterbuck is bad. This is the reason they don't fall prey to many animals or poachers. Its main enemy is the lion, although the young are attacked by cheetahs, leopards, and wild dogs.

On dry land the waterbuck is slow, and prefers to trot. But when excited or alarmed, it breaks into a gallop. It can move over broken or stony country with ease.

LILAC-BREASTED ROLLER

We also had a visit from the beautiful lilac-breasted roller. When clearly seen, the lilac breast and blue wings distinguish this bird from others. In flight it is one of the most beautiful of all birds, with its wings and tail showing different shades of blue and orange.

The lilac-breasted roller has a strong bill and thin pin-feathers on the outside of the tail. It is a rather noisy bird, drawing attention to itself with harsh calls. It can do the most remarkable aerobatics. It sits on a perch and surveys the surrounding countryside. When it spots an insect on the ground, it drops onto it with a characteristic wing action. The wings are held high above the neck when flying.

Lilac-breasted rollers live in pairs. But small family parties may also be seen. Large numbers gather at bushfires, where they catch insects in the air. They nest in the hole of a tree, with little attempt at disguise.

That day we also saw the colorful bee-eater, the malachite kingfisher, the gray heron, and the huge saddle-bill stork. Every now and then we saw a lily-trotter. It looks for insects under the large waterlilies. Its long toes enable the foot to cover a large area, so that the lily-trotter can walk on the minimum amount of floating material.

We saw baby crocodiles lying on the sandbanks with their mouths wide open. This enabled the herons to take pieces of food from between their teeth. As they heard our boat motors, the herons flew away. The crocodiles snapped their jaws shut and disappeared under the water.

Our presence alarmed a school of hippos which protested with loud snorts and blew jets of water at us. Then their big glistening bodies submerged into the water.

This area is made up of many lagoons. Light green water plants with shady trees cover the area. Impala, zebra and many other animals come to drink here.

The lilac-breasted roller is the most colorful bird in all of Africa.
It is also a remarkable acrobat when in flight.

The waterways twist and turn among sand channels down to the center, providing a life-giving stream of fresh water. Every stretch of water contains crocodiles, basking on the sandbanks, which they share with sandpipers, blacksmith plovers and many other birds.

After filming for some hours, we decided to rest awhile under the shade of a big mopani tree. A short distance away we saw a sable antelope making its way carefully to the water. It seemed to sense something and flung its head high. Something was not right. Further along the water's edge, hippos were enjoying themselves in the mud. There was nothing else to be seen. The sable went closer to the water, spurred on by its thirst. About a few yards from the drinking place it seemed very nervous. It bent its head to the water, with the curved horns almost touching its shoulders. Then it began to drink. Suddenly, a dark shadow appeared in the water. The sable panicked, but it was not quick enough. A crocodile emerged from the water and caught the sable with its jaws. Peter jumped up, wanting to help. But I pulled him back down. The sable had already been wounded. Even if we would have saved it, the sable would have been attacked by lions or wild dogs. One has to remember the law of the wilderness; the crocodile had to live on something. Slowly the crocodile pulled its prey into deep water.

The jackal most commonly found in Botswana is the black-backed jackal. It can be seen throughout all of southern Africa. It has a soft, fine coat, which is very tough. The white-striped back is sharply defined from the yellowish legs. Its head is like a fox. It has yellow eyes with a keen intelligent look.

THE BLACK-BACKED JACKAL

The black-backed jackal is 36 inches (91 centimeters) in length. The bushy tail is another 15 inches (38 centimeters). It stands 16 inches (41 centimeters) at the shoulders and weighs about 33 pounds (15 kilograms). They live in pairs but gather in large numbers at a kill. They usually eat animals that have been killed by lions. They also eat reptiles, insects, birds, and their eggs. They use their great speed to attack pythons. The jackal waits for its opportunity, then springs at the python taking a bite out of it. Then it backs off and waits for the next turn. It will also grab small snakes, dragging them along at full speed, preventing the snake to get a hold of the jackal. But a python looking for food can swallow a jackal whole.

The litter of jackals is about six pups. The young seem to be left to themselves at a very early age. I have often seen jackals less than half-grown trying to find food on their own. As soon as they leave the den, the mother does not trouble much with them. This may be the reason why many pups fail to survive.

Black-backed jackals do not appear to be dependent on water. They are often seen in very dry parts of Botswana. The variety of sounds which these colorful creatures make includes a harsh bark, and a yapping noise. The call of the jackal on a pitch-dark night in Africa has sent many a cold shiver down the spine of those spending their first night in the bush. At the same time it adds to the adventure. It becomes a night symphony, pleasing to the ears of the experienced bush-lover.

Aerobatics - spectacular flying feats.

Africa - a continent (large body of land) south of the Mediterranean Sea between the Atlantic and Indian Ocean.

Amphibian - an animal able to live on both water and land.

Antelope - a swift-running animal resembling a deer, found especially in Africa.

Baboon - a large African monkey.

Botswana - a country in southeastern Africa.

Bull - the male gender of certain animals.

Bushmen - a member of an aboriginal tribe of southern Africa.

Cheetah - a kind of leopard. Of the cat family, found in Africa.

Crocodile - a large reptile with thick skin, a long tail, and huge jaws.

Delta - a triangular patch of land at the mouth of a river.

Elephant - a very large land animal with a trunk and long curved ivory tusks.

Expedition - a journey for a particular purpose.

Feline - an animal of the cat family.

Heron - a long-legged, long-necked, wading bird living in marshy places.

Hippopotamus - a large African river animal with tusks, short legs, and thick skin.

Impala - a small antelope of southern Africa.

Jackal - a wild flesh-eating animal of Africa and Asia, related to the dog.

Kudu - a type of antelope found in Africa.

Lagoon - a small body of water connected with a large body of water.

Leopard - a large African and Asian flesh-eating animal of the cat family.

Lion - a large, powerful African and Asian flesh-eating animal of the cat family.

Makorro - a canoe made of wood by the African bushmen.

Mopani Tree - a tree found in the desert regions of southern Africa.

Okavango Delta - an area of the country of Botswana which is plentiful in water and wildlife.

Plain - a large area of level ground.

Poacher - a person who takes illegally from land or water.

Prey - an animal that is hunted or killed by another for food.

Python - a large snake that crushes its prey.

Roan - a horse-like animal with white and gray hairs.

Sable - a large antelope-like animal.

Sandpiper - birds with long pointed bills living in wet, sandy places.

Talons - the claws on a bird of prey.

Warthog - a kind of African pig with two large tusks and warts on its face.

Waterbuck - a type of antelope.

Wild dogs - wolf-like flesh-eating animals found in Africa; a member of the dog family.

Zebra - an African animal of the horse family, covered with black and white stripes.

INDEX:

Driftwood Library
of Lincoln City
801 S.W. Highway 101, #201
Lincoln City, Oregon 97367